Bruno Bartoletti

Unusual and strange animals in the World

The life of the most incredible and particular creatures that populate our planet

Premise

Welcome to the world of the strangest and most unusual animals! This book is a journey through the diversity and wonder of nature, discovering creatures that seem to come out of a fantasy world.

From animals that have adapted to living in the most extreme conditions on Earth, to creatures that have developed extraordinary abilities to hunt or defend themselves, to animals that seem to belong to another era or planet, there are plenty of examples of strange and unusual animals in this book.

But not only will you discover the amazing variety of shapes and colors of the most unusual animals, you will also learn how these creatures have evolved and how they interact with their surroundings.

Whether you are passionate about animals or simply curious to discover something new, this book will take you on an exciting and inspiring journey to discover the strangest and most surprising animals on our planet.

The Narwhal

The narwhal, known scientifically as Monodon monoceros, is a cetacean of the monodontid or dolphinater family. Its name is derived from the Norwegian "narhval," which means "dead whale", probably because of its body-like appearance when in the water.

The narwhal is similar to the beluga, but it has a unique feature: a spiral tooth that develops in the upper jaw. The tooth has a right-to-left winding and is usually only present in males, although in rare cases it may also be present in females. The male's tong can reach the length of 2.4-2.7 meters and can be used for different functions, such as competition between males during the mating season or as a tool to look for food on the seabed.

The narwhal can reach a length of about 4-5 meters, excluding the tanch. It has a rounded head, a dorsal fin about 4-5 cm high and between 60 and 90 cm long, and small, round swimming fins. Adults are greyish-white with blackish spots on their backs, while young people are typically darker. Older specimens can take on an almost white color.

The narwhal lives mainly in the Arctic and sub-Arctic and rarely moves away from this area. The southern limit of its spread area roams around the 70th North, but sometimes moves south and can run aground on the coasts of Britain and the

Netherlands. However, due to excessive hunting and climate change, the survival of this species is at risk.

Usually narvals live in small groups, consisting of about 50 units, but they can associate to form much larger groups, composed of several thousand individuals, both mixed and divided by sex. The narwhals swim quite fast and, when they rise to the surface, they emit a high-pitched whistle and remain motionless on the surface of the water for a few minutes before diving again.

During the summer, narvals move towards the bays and sometimes go up the rivers, as evidenced by the discovery of a specimen about 1000 km from the mouth of the Yukon River.However, they sometimes get trapped in the bays, obstructed by the ice progressively covering the entire inlet. In these situations, the Narwals try to open large holes to breathe and, when caught in groups, provide the Inuit with meat and fat for the winter, because the "sacssats," as the Inuit call these "cells," can hold up to 1000 narvals each.

Narvals feed mainly on cuttlefish, cephalopod molluscs and crustaceans, but have no functional teeth. Instead, they grab the prey with the solid ends of their mighty jaws and swallow it whole.

Like many other cetacean species, the reproductive habits of narvals are little known. The little data available is only obtained when the animals are kept in captivity in the 'oceanaries'. It is known that the female births one or two young, which measure about 1.5 m at birth. According to a Greenlandic legend, the baby's tail comes out of the mother's body four to six weeks before giving birth. This mention has also been reported in some scientific texts, but to prove its veracity, further more detailed and in-depth observations, so far limited, would be needed.

The Vikings and the Siberian populations brought the fangs of the narwhal to Europe, which became an important source of wealth for Viking Iceland. This product was in high demand throughout Europe and many works of art such as St. Mark's Basilica in Venice still retain these tusks today. In the Middle Ages, narwhal tusks were marketed as fine ivory and as a drug to neutralize poisons and increase male sexual potency. This fueled the legend of the unicorn, as the twisted tooth of the narwhal looks very much like the image of the unicorn. However, it is likely that this legend was created by merchants to increase the attractiveness and price of tusks. The Imperial Treasury of Vienna also preserves a large narwhal fang, a gift from King Sigismund II of Poland to Emperor Ferdinand I of Habsburg, incorporated into the Habsburg treasury as a "unicorn horn" (Ainkhürn). Solo verso la metà del XVII secolo si diffuse la consapevolezza della vera natura di questi reperti.

Photo below: Fang of Narvalo exhibited in the Doge's Palace in Mantua Italy

The Ornitorinco

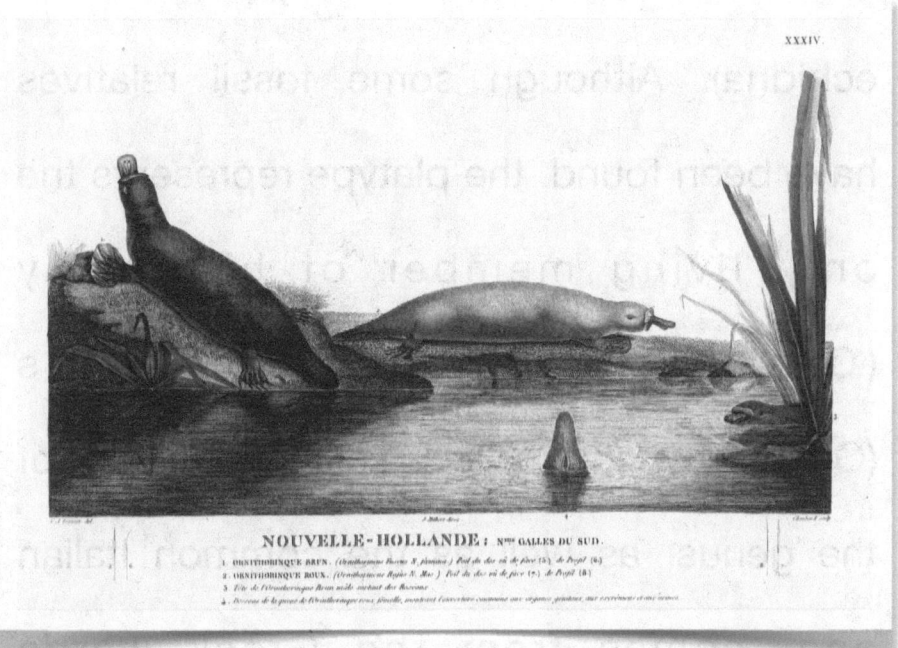

NOUVELLE-HOLLANDE : N°ᵉˢ GALLES DU SUD.

The platyp, also known as the platypus (Ornithorhynchus anatinus Shaw, 1799), is a small semi-aquatic mammal endemic to eastern Australia. It is part of the order of monotremes, which includes only five species that still exist, all characterized by spawning instead of the birth of young (the other four are collectively known as echidna). Although some fossil relatives have been found, the platype represents the only living member of his family (Ornithorhynchidae) and genus (Ornithorhynchus). The scientific name of the genus, as well as the common Italian one, comes from the Greek words "òrnis" (bird) and "rhýnchos" (muse). The second term of the scientific name of the species, on the other hand, comes from the Latin term for "anas" (duck).

For a long time the platypus and other monotrems were poorly understood and still today some of the 19th century legends surrounding them persist, especially in the northern hemisphere. For example, some people still believe that monotremes are "inferior" or almost reptiles and that they are the distant ancestors of "superior" placental mammals. However, today it is known that modern monotremes represent the oldest branch of the mammal family tree, with a subsequent branching that led to the marsupial and placental groups. The oldest fossils of monotrems, such as Teinolophos and Steropodon, are closely related to the modern platypus. In summary, the ornithorinch is one of the closest relatives of ancestral mammals, but it does not represent a direct link in the evolutionary chain of mammals. Instead, it is a fairly distinct branch from all the other known ones.

In 2004, Frank Grützner and Jenny Graves of the Australian National University of Canberra discovered that the platypus has a sex chromosome system consisting of 10 chromosomes, unlike the 2 (XY) found in most other mammals (with the exception of the Belize howler monkey). In detail, males possess 5 pairs of X chromosomes and 5 pairs of Y chromosomes in sequence XYXYXYXYXY, while in females the sequence is made up of 10 pairs of X chromosomes (XXXXXXXXXX).

The platypus chromosome system has some characteristics typical of mammals, but also some similarities with the WZ system of birds. For example, the DMRT1 gene, which is believed to be involved in determining the sex of birds, is present in the platypus in a very similar form, and the DM domain of the corresponding protein is highly preserved in vertebrates. This domain also regulates sexual differentiation in insects and worms, suggesting an ancestral function in sexual differentiation much prior to the separation between protostoms and deuterostoms. This function has been lost in many evolutionary branches, but in some cases it has re-emerged with similar or very different dynamics, such as in birds (probably already present in the common ancestor of ornithodirs) and mammals (probably already present in the common ancestor of therapsids).

The physiology of the platypus is unique compared to other mammals.

Its metabolic rhythm is considerably lower, with an average body temperature of 32°C instead of the 38°C typical of placental mammals or Eutera.

It is not yet clear whether this is a characteristic of monotremes or an adaptation of the few species that have survived difficult environmental conditions.

The platypus has a body and a wide flat tail covered in brown fur.

It has palmate feet and a wide snout, as hard as rubber, which resembles more that of a duck than that of any other known mammal. Precisely because of these characteristics it is known as a duck-billed platypus or Duck-billed Platypus. During the colonial period, however, it was called Water Mole.

The size of the platypus varies greatly, from less than a kilo to more than two kilos, with a body length between 30 and 40 cm and that of the tail between 10 and 15 cm.

Males are about a third larger than females. Curiously, there doesn't seem to be a climate rule that explains regional variations in size.

Puppies have three-cusped tribosphenic molars, which are one of the distinctive features of mammals, while adults are toothless.

The jaw/maw is different from that of other mammals and the muscle that opens it has a different structure.

The ossicins that carry the sound in the inner ear are completely embedded in the skull, as in all real mammals, instead of being found in the jaw as in Cynodontia and other pre-mammalary synapses.

However, the external opening of the ear is still at the base of the jaw. The ornithorinch has additional bones in the scapular belt, including an absent interclavicle in other mammals.

It also has a reptile gait, with legs placed on the sides of the body instead of below it.

Finally, like all monotrems, the platypus has the cloaca, or a single orifice from which to lay eggs, urinate and defecate, a characteristic that it shares with fish, birds and other monotremes.

The male of the platypus is equipped with a hollow spur in each of his hind legs, which he uses as a defensive weapon against predators or in fights for the territory.

It is not certain whether the use of the spur is exclusively related to the defense, since this characteristic is missing in the female of the platypus.

Unlike poisons produced by other non-mammal species, spur venom appears to contain molecules with a molecular weight of less than 5,000 daltons, such as peptides consisting of a few amino acids joined together by peptide or carboamide binding, or other molecules that are not deadly but can cause serious injury to the victim.

There is still no antidote for the poison of the ornithorinch, which causes in human subjects a strong and immediate pain around the wound, accompanied by edema that gradually extends to the affected limb. Anamnestic and anecdotal studies have shown that pain can persist for days or even months in the form of hyperalgesia.

Although the poison is not lethal to humans, it can instead pose a deadly threat to dogs and small pets.

The platypus lives in semi-aquatic and nocturnal habitats, ranging from the cold of the mountainous regions of Tasmania and the Australian Alps to the tropical rainforests of the Queensland coast to the north to the base of the Cape York Peninsula.

Its inland distribution is not well known, but it is extinct in southern Australia and is no longer in the main part of the Murray-Darling basin due to the declining water quality.

Its distribution along coastal river systems is unpredictable, but it seems to be present even in degraded rivers.

The platypus is a great swimmer who spends a lot of time in the water, with his eyes completely closed and all four limbs palmed. He eats worms, insect larvae and freshwater shrimp, which he catches by digging into the river bed with his muse or swimming. It is one of the few mammals to possess a highly sensitive sense of electrolocation to locate prey, as well as a very sensitive beak.

When not in the water, the platypus retreats into a short oval den in the embankment, often hidden under a tangle of roots.

During the breeding season, males fight for females using their poisonous spurs.

The female lays her eggs in a nest she has dug, up to 20 meters long and locked at intervals with caps, while the mother secretes milk from the pores of the skin for her fur-free puppies.

The blobfish

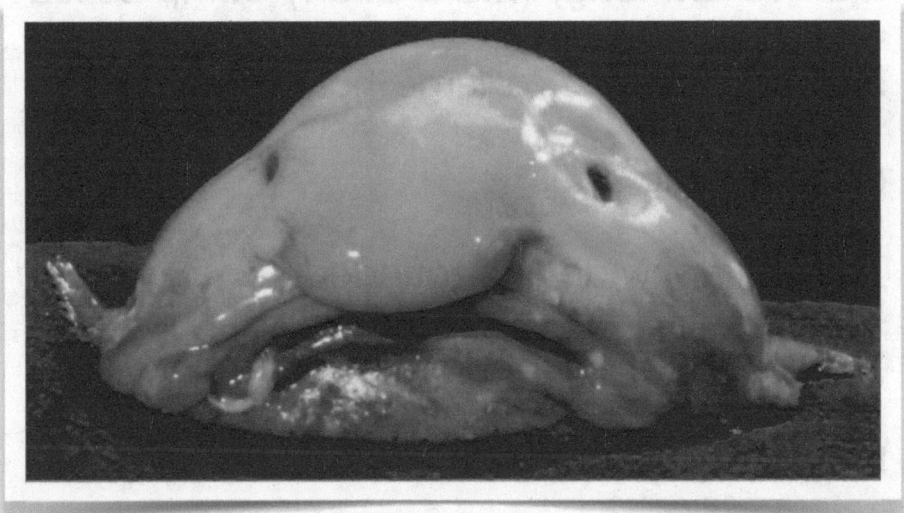

The blob fish, belonging to the family Psychrolutidae, is known scientifically as Psychrolutes marcidus (McCulloch, 1926). This abyssal animal is characterized by its soft conformation and for this reason it is commonly called "blob fish".

The Psychrolutes marcidus is usually less than 30 cm long, with a poorly compressed body at the hips, a large head and eyes of considerable size.

The fins are wide and rounded.

The color of his body is gray-pink, with brown spots. The mouth and lips are pinkish-white in color.

Due to the high pressures at which it lives, which can be 60 to 120 times greater than those perceived on the surface, the use of a swimming bladder to maintain the hydrostatic state is not effective.

However, the meat of Psychrolutes marcidus has a gelatinous appearance, with a specific gravity slightly lower than that of water, which allows it to float on the ocean floor without having to swim and without consuming energy.

Its relative lack of muscle does not represent a disadvantage, as it mainly ingests edible floating material, such as crustaceans found in the ocean depths.

The common impression that Psychrolutes marcidus is a deformed being is partly due to the decompression trauma suffered by the first captured specimens.

In fact, in a natural environment, it looks very similar to that of other more familiar boney fish.

Despite this, he was voted the 'ugliest animal in the world' and named a mascot of the Ugly Animal Preservation Society.

The body of Psychrolutes marcidus, as mentioned, has a gelatinous consistency and a higher density than that of water, which allows it to live in great depth and float on the seabed.

For all intents and purposes, this animal is devoid of musculature, and is only equipped with two small fins that allow it to orient itself but not to swim like other fish.

Despite its reduced mobility, the Psychrolutes marcidus lives quietly on the bottom and lets itself be carried away by the winds and currents without having to make any effort to hunt.

The blob fish has a sad air because of its downward-facing mouth and large triangular head, so much so that it deserves the nickname of "sad-faced fish".

Its typical, grayish-pink color, does not improve its appearance, making it unpleasant to the eye.

This species is inedible because of its appearance and consistency, which would be unpleasant to the palate, and it often ends up in fishermen's nets, thus risking disappearing from the seas of New Zealand and Tasmania.

The Colibrì Ape

Immagine di John Gould - Gould, John: A monograph of the Trochilidae, or family of humming-birds Volume 3 (1861)

Mellisuga helenae is a species of bird that lives exclusively on the island of Cuba.

Despite being part of the hummingbird family, he is incredibly small, even compared to other representatives of the same family.

With a length of about 5 centimeters from the tip of the beak to the end of the tail and a weight of about 1.8 grams, it weighs less than a penny coin.

This species of hummingbird would be ideal for inspiring stories and fairy tales, not only for its tiny size, but also for other aspects of its life.

For example, the female builds a nest only 3 centimeters in diameter and her eggs are the size of a pea.

It is a bird so small that it is often confused with a bee.

This is why in Cuba he received the nickname of 'zunzuncito' because of the characteristic hum of his wings during the flight.

Adult males of this species have a deep red head and throat, with similar blashes extending along the sides of the chest.

In addition, both the head and neck have iraging reflections ranging from ruby red to orange, depending on the surrounding light.

The male Hairlop Ape have the upper parts of a tale green color, while the lower parts are white with green hues on the sides.

The tail, serrated, is blue with black traces at the corners.

An interesting fact is that young males, at first glance, resemble females, but are smaller and do not have the characteristic white-pointed tail.

Instead, although females have a resemblance to adult males, they are a little older and do not have the iridescent "helet" typical of males.

In addition, the tips of the feathers on their tail are white.

His flight skills are extremely interesting and need to be deepened.

In fact, it is known that these hummingbirds spend much of their lives in flight, more than any other species.

Their legs are only used to perch if necessary.

For this reason, the flying muscles of Mellisuga helenae have evolved in a particular way, making up 35 % of their total body weight.

In addition, these birds have located, thanks to the continuous evolution of the species, these large muscles right on the sternum that is the fulcrum to make the wings float in the air and have developed the same in conical shape so that they can fly efficiently, without a great expenditure of energy, as is also the case with other members of the hummingbird family.

It is true that hummingbirds are able to move their wings according to an eight-shaped pattern, which allows them to remain stationary in the air, a capacity called stationary flight.

This is possible thanks to the unique anatomy of their wings and joints.

The wings of hummingbirds are very long relative to their bodies and are made up of thin, light bones.

The pens are arranged to form a rigid, but flexible flight surface. In addition, the wings of hummingbirds are able to rotate 180 degrees thanks to the structure of their shoulder joint, which allows them to move independently of the rest of the body.

The eight-shaped movement of the wings is generated by the powerful pectoral muscles of the hummingbird which, as mentioned before, occupy a significant part of their body mass.

These muscles contract quickly and generate the force needed to move the wings precisely and quickly, allowing the hummingbird to fly forward, backward and even remain stationary in the air.

In summary, the unique anatomy of the wings and joints of hummingbirds allows them to fly in an extraordinary way, moving quickly and accurately both forward and backward, and to remain stationary in the air, making them one of the most incredible birds in nature.

In the research done on the way of communicating of the Hummingbirds in general, it must be said that what has been learned so far is not entirely correct for the Mellisuga helenae, as it is a kind of hummingbird in which only males sing to attract females during the mating season.

Females don't produce songs.

In addition, the sounds produced by the males of the Mellisuga helenae are actually complex and distinct, although they may seem simple and unattractive to the human ear.

Their songs include a variety of sounds, including trills, whistles and chirping, which can only be heard thanks to the use of special instruments such as particular types of microphones.

In addition, Mellisuga helenae male songs often include a single note repeated many times, but each note can last even more than a second, unlike what researchers in general say about the Humbird species.

They are distinguished by their ability to produce a wide range of sounds, from melodious songs to noisy cries. Their songs are often used for communication and courtship during the mating season, but also for territory and alert to predators.

Foto: Charles J. Sharp

The Pangolino

Foto: Piekfrosch

The pangolin is a nocturnal animal belonging to the order Pholidota, which includes eight different species.

It is mainly found in Africa and Asia, where it lives in forests and savannahs.

The pangolin is known for its skin covered with flakes, which protect it from predators. These scales are made of keratin, the same substance that forms nails and human hair. When he feels threatened, the pangolin rolls up in a ball to protect himself, exposing his scales like an armor.

The pangolin is an insectivorous animal and feeds mainly on ants and termites.

Its long, sticky tongue allows it to catch prey within their nests.

Unfortunately, the pangolin is an endangered species due to the illegal hunting and trade of its scales, which are used in some parts of the world for medicinal purposes.

There are also concerns about the loss of the pangolin's natural habitat due to deforestation and agricultural expansion.

However, there are ongoing efforts to protect the pangolin and its species from hunting and illegal trade, through the implementation of laws and public awareness of its importance to the ecosystem.

Ancient representation of a pangolin in a table of the Acta Eruditorum of 1689

The Limulo

The lime, also known as Atlantic horseshoe crab or king crab, is not closely related to crabs, but they share a closer relationship with arachnids, such as spiders, ticks and scorpions.

The scientific name of the limulus is Limulus polyphemus, although in the past it has been identified as Limulus cyclops, Xiphosura americana or Polyphemus occidentalis.

It is a chelicerate arthropod and represents the only member of the genus Limulus.

The simuli feed mainly on molluscs and marine worms that are found on the seabed. These particular animals live in North America and the Gulf of Mexico, and make a massive migration during the spring to reach

Delaware Bay, where they lay their eggs. Females produce a number of eggs ranging from 15,000 to 64,000 depending on their size, before mating. The eggs are stored in dense masses at the front of the carapace, and are laid in 4-5 holes 15-20 cm deep dug into the sand during each tide. Each brood contains about 4000 eggs, and a female lays about 20 broods in a year over the course of multiple tides.

The limuli live in coastal areas with depths ranging from 5 to 50 meters, although they can move to deeper waters in search of food.

The immune system of the Limulus has a primitive ability to recognize lipopolysaccharides on Gram-negative bacteria and to eliminate them by coagulation. This capability led to the creation of the in vitro LAL test, also known as the Limulus Test, commonly used to detect bacterial endotoxins in industrial raw materials, water, drugs, and some bacterial diseases.

However, pharmaceutical companies catch the Limuli to recover their blood, then then throw them into the sea after picking it up.

500,000 Limuli are collected each year along the east coast of the United States.

Limuli are pierced by needles in the heart area to drain 30% of their blood, which is worth about $18,000 per liter. From this blood is obtained an amebocyte lysate, which contains the enzymes that allow gelling in the presence of bacterial endotoxin (LPS). The blood of the Limuli is almost transparent, but turns blue in contact with air due to the oxidation of hemocyanin, a copper-containing respiratory pigment that has the same oxygen transport functions as hemoglobin.

In addition, several peptides isolated from Limulo hemocytes have been shown to inhibit HIV proliferation in in vitro tests.

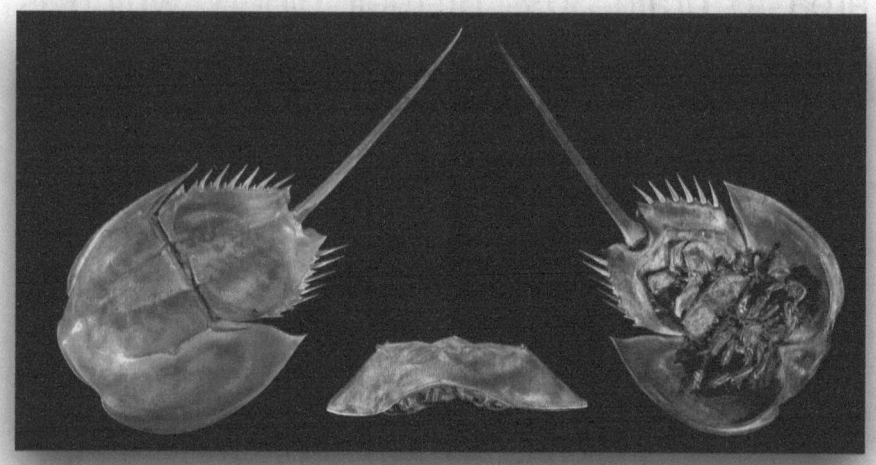

Foto: Didier Descouens

The body of the limuli is composed of three distinct parts. The first part is the head, also known as the prosoma or cephalothorax, which has a characteristic horseshoe shape. In this section of the body are the eyes, brain, heart, mouth, chelicera and locomotor legs. The second part is the chest, also known as opistosoma or abdomen. This section of the body features side thorns and encloses five pairs of book gills, a genital operculum, and an opening of the genital pores. The third part of the body is the tail, known as telson, which is long and stiff.

The shell of limuli can range from greenish gray to dark brown, while adults can reach a maximum length of 60 cm. The silt is equipped with six pairs of appendages, the first two of which are the cheliceri, smaller and claw-shaped. The second pair of appendages is known as pedipalps, which in males is modified to end in a claw that allows them to cling to females during mating.

The three pairs of ambulacral legs are used to break up food and bring it to the mouth, located in the center of the cephalothorax and devoid of jaws. The last pair of motor legs is adapted to allow locomotion on the sandy ground of the beaches, with tarsal spurs that open as an umbrella and are also used to dig in the mud. In the center of the abdomen are the chilari, small and unarticulated appendages that correspond to the claws present on the three pairs of

legs.

The limul has six pairs of gills per book, the

first pair, known as the opercolo, is larger in

size and contains the opening of the genital

pores. The other five pairs of respiratory

organs are covered by the operculum. The

gill appendages are also used for swimming

and carry chemosensors sensitive to the

characteristics of water. The silt can also

breathe on land for a short period of time, as

long as the gills remain moist.

The tail of the limulus is used to be able to

turn in case of overturning and is equipped

with photoreceptors that intervene in the regulation of the circadian rhythm.

The males of the limuli differ from the females in their size, 20% smaller, and by their modified pedipalps that allow them to hold on to their companions during mating to fertilize the eggs. In addition, the front of the carapace is wider in females than in males.

The limulo is a very important organism in the study of the physiology of vision, thanks to its complex visual structure.

It has two compound eyes located on the top of its shell and five simple eyes that are sensitive to light. Among them, there are two median eyes that can also perceive ultraviolet light and two rudimentary side eyes that vary their sensitivity based on the signals sent by the brain in relation to an internal clock.

Compound eyes are an exception in the chelicerate family, as no other species possesses this type of eyes.

They are mainly used to locate a partner.

During the night, the view of the limulus is amplified by a million times thanks to a series of photoreceptors present on the top and sides of the telson, which send synchronization signals to the brain with the cycles of light and dark.

In addition, on the belly of the silo, there are two eyes that favor orientation while swimming. The two compound eyes, on the other hand, each have an ommatid connected to a very large nerve fiber, which allowed the cellular study of the nerve response to stimulation.

Thanks to these studies, conducted since the twenties of the twentieth century, it was possible to identify the mechanisms of operation of visual phenomena such as lateral inhibition, which makes it possible to distinguish lines, shapes and contours of objects.

The reproduction of sirolimuss takes place in the spring, when particularly high tides occur during the full moons. Before mating, females produce 15,000 to 64,000 eggs, depending on their size, and store them in dense masses at the front of the carapace. During each tide, females dig 4-5 holes 15 to 20 cm deep in the sand to lay their eggs. Each brood contains about 4000 eggs, and during the course of several tides, a female lays about 20 broods per year.

The Ghost Tarsius

The Tarsium Spectro, also known as Maki

Folletto and scientifically referred to as

"Tarsius tarsier" is a species of primate belonging to the family Tarsiidae.

This species represents the type of belonging for this family.

What distinguishes it from other primates is the fact that it feeds exclusively on meat, making it a pure carnivore.

The Tarsium Spectrum is of great importance to scholars because it is believed to represent the missing link between the proscimmia and the actual monkeys, as in the past it had been classified as part of the former. In addition, this species was used as a "container" in which a large part of the Tarsiidae were included as a subspecies.

However, thanks to the work of the scholars, four new species have been classified (Tarsius dentatus from T. spectrum dentatus, Tarsius pelengensis from T. spectrum pelengensis, Tarsius pumilus from T. spectrum pumilus, Tarsius sangirensis from T. spectrum sangirensis), while Tarsius spectrum has become synonymous with Tarsius tarsier.

The coat of the tarsio is gray and velvety, while its tail is flaky like that of a rat, but it has a tuft of dark hair on the tip. The total length of the animal is about 35 cm, two-thirds of which consists of the long tail, and the total weight is about 120 g.

The eyes of the tarsium are huge, with the eye socket exceeding the size of the stomach and brain.

This is due to the fact that tarsium is a nocturnal animal that lacks tapetum lucidum, a reflective membrane inside the eye.

The eyes are fixed in the orbits, but the tarsio can rotate the head by more than 180° to look around.

The ears of the tarsium are similar to teaspoons and rest on short tubular handles.

The head and body together reach just half the length of the hind legs, which are made up of thighs, shins and feet, all very elongated and of similar length.

The word "tarsio" comes from the elongated tarsus. The animal's tibia and fibula are fused at the end to absorb the impact when the tarsium jumps from branch to branch.

The hands of the tarsio have very elongated fingers and swollen fingertips for a secure grip on almost smooth surfaces. However, adhesive fingertips are less evolved than other species such as Carlylet syrichta. The index and middle fingers are equipped with special claws for grooming activity (or hair cleaning), while the other fingers have nails similar to human nails.

We are talking about an animal with nocturnal and twilight habits.
During the day, he rests upright between the leaves, while at sunset he wakes up and begins to clean his hair.

After grooming, the animal ventures in search of prey until the morning. Thanks to its long hind legs, it is able to jump up to 6 meters away. During the jump, the animal rotates its front legs first and then its hind legs, using the tail as a barbell. On the ground, the animal moves by jumping sideways, holding its tail vertically, similar to the Malagasy sifaka.

He lives alone or in small family groups, each with their own territory that is defended with urine and glandular secretions. The tarsium has an epigastric gland at the level of the chest. The members of the group communicate through vocalizations, and often the male and female perform in duets to demarcate their territories. Clashes are rare due to the marking of boundaries with urine and glandular secretions.

The Ghost Tarsium or Spectre Tarsium, as it is called, lives mainly in the lowlands of the islands of Sulawesi and Selayar, Indonesia. It is often found in areas of secondary rainforest, which was cut to the ground for some reason and then started growing again. It is believed that this preference is due to the greater presence of bamboo, bushes and tall grass in this type of forest compared to the primary type.

The tarsio is venerated as a totem animal by the tribes of Iban headhunters in the areas where it lives. It is believed that the animal can detach its head from the body, thanks to its ability to twist the neck up to 180 degrees. The Ibans believe that calling the tarsio by his name is taboo, as this could attract his anger to the offender's house.

Other Sulawesi ethnic groups, on the other hand, keep this animal in captivity for its gentle nature and intelligence. However, the breeding of tarsium can be very problematic as it requires live food and can die in a few days in the absence of it. Also, if you catch a tarsio separating it from its partner, both could let themselves die of starve. In general, the tarsio is a very sensitive animal and the trauma of the capture can scare him to such an extent that it drives him crazy and makes him hit his head against a wall or cage bars.

The Electric Eel

Foto di Steven G. Johnson

The electrophore, or Electrophorus electricus, is a freshwater fish native to South America belonging to the Gymnotidae family. Also known as an electric eel or electric gymnoto in the past, he is famous for his ability to generate powerful electric fields through modified muscles located along his hips.

These electric fields can reach a potential difference of the order of a few hundred volts and are used for both hunting and self-defense.

Despite its common name, the electrophore is not an eel, but a knife fish.

Until 2019, the electrophore was the only species of its kind.

The electric eel is equipped with three pairs of abdominal organs that allow it to generate electricity: the main organ, Hunter's organ, and Sach's organ.

These organs make up most of his body and produce two types of electrical discharges: low voltage and high voltage.

The electrolytes that compose them are aligned in such a way that a current of ions can flow through them and stacked in such a way as to generate a difference in electrical potential.

When the electric eel detects its prey, the brain sends a signal through the nervous system to the electrolytes that activate producing a sudden difference in electrical potential, generating an electric current similar to that produced by a battery.

In this way, the electric eel can control the nervous system and muscles of the prey via electrical impulses, preventing it from escaping or forcing it to move to locate its position.

The electric eel can produce a discharge of up to 860 volts and 1 ampere of current, making it the most powerful and dangerous electrophore fish, followed by electric catfish and torpedoes.

Low voltages are used to detect the surrounding environment, while high voltages are used to detect and stun prey.

Sach's organ, which is located inside the electric eel, is associated with electrolocalization, a mechanism that allows it to locate prey and communicate with other specimens of its species.

Young people produce lower voltages and can vary the intensity of the electric discharge depending on the situation.

The electric eel has a long, cylindrical body that can grow up to about 2 meters in length and weigh up to 20 kg, making it the largest species among the Gymnotiformes. Its typical coloration is dark gray-brown on the back and yellow-orange on the belly, with mature females having a darker livery on the abdomen. Unlike many other fish, the body of the electric eel is not covered with scales and the mouth is positioned at the end of the muzzle and has a square shape. In addition, it does not have a dorsal fin, but its anal fin extends the entire length of the body to the tip of the tail.

Like many other ostariophysial fish, the electric eel has a swim bladder with two chambers. The anterior chamber is connected to the inner ear through a Weber apparatus, composed of small bones derived from the cervical vertebrae, which improves its auditory ability. The rear chamber extends along the entire length of the body and helps the fish maintain buoyancy.

The electric eel is an obligated aerobic fish, which means it requires oxygen to survive. For this reason, it possesses a vascularized respiratory system that allows it to exchange gas through the epithelial tissue in its buccal cavity. The animal goes up to the surface to breathe every ten minutes or so, collecting almost 80% of the oxygen used by the fish.

If you want to write to the author you can do

it on s1er050@yahoo.com